I0469332

Three Firefighter Fatalities
in Training Exercise
Milford, Michigan

Investigated by: J. Gordon Routley

This is Report 015 of the Major Fires Investigation Project conducted by TriData Corporation under contract EMW-86-C-2277 to the United States Fire Administration, Federal Emergency Management Agency.

Department of Homeland Security
United States Fire Administration
National Fire Data Center

U.S. Fire Administration Fire Investigations Program

The U.S. Fire Administration develops reports on selected major fires throughout the country. The fires usually involve multiple deaths or a large loss of property. But the primary criterion for deciding to do a report is whether it will result in significant "lessons learned." In some cases these lessons bring to light new knowledge about fire--the effect of building construction or contents, human behavior in fire, etc. In other cases, the lessons are not new but are serious enough to highlight once again, with yet another fire tragedy report. In some cases, special reports are developed to discuss events, drills, or new technologies which are of interest to the fire service.

The reports are sent to fire magazines and are distributed at National and Regional fire meetings. The International Association of Fire Chiefs assists the USFA in disseminating the findings throughout the fire service. On a continuing basis the reports are available on request from the USFA; announcements of their availability are published widely in fire journals and newsletters.

This body of work provides detailed information on the nature of the fire problem for policymakers who must decide on allocations of resources between fire and other pressing problems, and within the fire service to improve codes and code enforcement, training, public fire education, building technology, and other related areas.

The Fire Administration, which has no regulatory authority, sends an experienced fire investigator into a community after a major incident only after having conferred with the local fire authorities to insure that the assistance and presence of the USFA would be supportive and would in no way interfere with any review of the incident they are themselves conducting. The intent is not to arrive during the event or even immediately after, but rather after the dust settles, so that a complete and objective review of all the important aspects of the incident can be made. Local authorities review the USFA's report while it is in draft. The USFA investigator or team is available to local authorities should they wish to request technical assistance for their own investigation.

This report and its recommendations were developed by USFA staff and by TriData Corporation, Arlington, Virginia, its staff and consultants, who are under contract to assist the Fire Administration in carrying out the Fire Reports Program.

The U.S. Fire Administration greatly appreciates the cooperation received from the National Transportation Safety Board, the Catlettt Volunteer Fire Company, the Fauquier County Office of Emergency Services, the Fauquier Fire/Rescue Association, and the National Railroad Passenger Corporation.

For additional copies of this report write to the U.S. Fire Administration, 16825 South Seton Avenue, Emmitsburg, Maryland 21727. The report is available on the Administration's Web site at http://www.usfa.dhs.gov/

U.S. Fire Administration

Mission Statement

As an entity of the Department of Homeland Security, the mission of the USFA is to reduce life and economic losses due to fire and related emergencies, through leadership, advocacy, coordination, and support. We serve the Nation independently, in coordination with other Federal agencies, and in partnership with fire protection and emergency service communities. With a commitment to excellence, we provide public education, training, technology, and data initiatives.

TABLE OF CONTENTS

Three Firefighter Fatalities in Training Exercise
Milford, Michigan
October 25, 1987

Local Contacts: Chief James Frankfurth
Milford Fire Department
Milford, Michigan 48042
313-348--3444

Captain John Sura
State Fire Marshal Division
Michigan State Police Office
7150 Harris Drive
Lansing, Michigan 48913
517-332-5452

SUMMARY OF KEY ISSUES

Issues	Comments
Ignition Scenarios	Multiple training fires in old single family frame house set with large amounts of accelerants. One exit door.
Training Procedures	Not in compliance with NFPA 1403, Standard for Live Training Fires in Structures. Lacked several safeguards for a live training fire, including sufficient paths of escape, manning of safety lines, less flammable environment, and a safety officer.
Firefighter Training	Training on response to being trapped and training for emergency situations involving SCBA may have needed improvement.
Protective Garments	Firefighters killed were fully suited; garments appeared to provide adequate protection. Burns were not the problem.
Breathing Apparatus	Straps and certain other parts may have burned through or failed quickly, reducing time to escape.
Communications	Inadequate between those inside and those outside of training house.

OVERVIEW

An unusual training exercise involving simulated arson sets and live firefighting evolutions in an abandoned farmhouse resulted in the deaths of three volunteer firefighters and injuries to three others. The incident occurred in Milford Township, Michigan, a rural area approximately 30 miles from Detroit, on October 25, 1987. Four area volunteer departments participated in the exercise and the fatalities included members of three of the departments.

This incident reinforces many of the lessons on firefighting safety that have been learned in real fires and in several previous incidents at live fire training exercises. Only five days earlier, a firefighter died in a mishap while igniting a training fire in Hollandale, Minnesota. National Fire Protection Association (NFPA) records indicate an average of one death per year under similar circumstances, as well as numerous injuries. Live fire training accidents continue to be a source of unacceptable danger to the fire service in the United States.

This is believed to be the first multiple death training incident in the United States since the adoption of NFPA Standard 1403 in 1986, which specifies safety procedures for live fire training evolutions in structures. The development of this standard was prompted by earlier tragic incidents, but was delayed by controversy and disagreement over the need for such a document or the wisdom of its requirements. It is a consensus standard and there is no legal requirement for it to be followed in the State of Michigan.

The drill was arranged and directed by the Milford Fire Department, with the other departments participating as guests. It was intended to familiarize the firefighters with the evidence that would result from arson fires. To achieve this objective, several different arson sets were prepared in the two-story abandoned structure, using flammable and combustible liquids as accelerants.

The plan was to ignite all of the arson sets and then extinguish the fires, allowing the trainees to examine the evidence before and after the fact. Some of the trainees would gain added experience in interior structural firefighting and in the use of self-contained breathing apparatus (SCBA). A tanker shuttle was to be used to supply water for the drill, providing an opportunity for all four departments to practice this method, and members of an Explorer Scout group were on hand to practice operating exterior handlines.

The plan was to reignite the house and burn it to the ground after completion of the interior operations and examination of the evidence. Exterior 2-1/2 inch hoselines were positioned to protect an exposed new house under construction on the same property.

The deaths and injuries occurred because of an unanticipated flashover on the ground floor that trapped six members on the upper level. Three of those members escaped, but three were killed as the fire rapidly extended to the second floor. All three deaths were attributed to inhalation of products of combustion, although all were wearing full protective clothing and SCBA.

The dead firefighters included a 41-year-old female member of the Milford Township Volunteer Fire Department, a 34-year-old male member of the Lyon Township Volunteer Fire Department, and a 33-year-old male member of the Highland Township Volunteer Fire Department. The three injured firefighters were all members of the Milford department.

THE HOUSE

The exercise was conducted in an abandoned two-story frame farm house, reported to be over 100 years old, located approximately one mile southeast of the Town of Milford, in Milford Township. The property owner had donated the house to the fire department and plans called for it to be destroyed in a controlled burn after interior operations were completed.

The house included a one-story section that contained the kitchen and an entry/storage room, and a two-story section that included two rooms on the ground floor and two bedrooms and a corridor on the second floor. (See Appendix B for photographs and floor plans.) The construction was wood frame, originally with lath and plaster interior walls and ceilings. Both sections had peaked roofs with attic spaces.

The house appeared to have been modernized while still occupied, with new kitchen cabinets and appliances and the installation of wall paneling and ceiling tile throughout the ground floor level. The wood paneling and low density fiberboard ceiling tiles were attached to furring strips nailed into the original lath and plaster. The configuration of walls and doors on the ground floor had also been modified, leaving a door at the northeast corner of the single-story section as the only exit to the exterior.

Since being abandoned the house had fallen into a state of disrepair. There were holes in the ground floor walls and ceilings. On the second floor there was no ceiling in the bedroom and several holes in the corridor ceiling. Where a bathroom had previously existed, fixtures had been removed, leaving holes in the floor and walls, and the doorway separating the bathroom and corridor had been removed. This left the former bathroom as a part of the corridor, with a sink and tub sitting on the floor, along with accumulations of trash and debris. During preparations for the exercise, the bathroom sink was placed over a hole in the floor that could have presented a danger to personnel on the upper floor.

There was no railing around the top of the stairway opening. A wall at the ground floor level partially enclosed the stairway, but there was no door to stop the spread of fire or smoke from the ground floor up the stairs.

The stairway was the only exit from the second floor. To reach the exterior from the second floor, a person would have to go down the stairs, through the dining room and kitchen areas, and through the rear story area to the only available door. The travel distance from the bottom of the stairs to the exit door was approximately 35 feet.

There were six windows on the second floor, but only one provided access to the rooftop of the kitchen, and this window was immediately adjacent to the top of the stairs. Trees and shrubs partially obscured several of the windows and no ladders were placed to provide alternate exits prior to the drill.

PREPARATIONS FOR THE EXERCISE

The Assistant Fire Chief was responsible for planning the drill and establishing the arson scenarios. In preparing the arson sets, approximately 4 to 5 gallons of flammable and combustible liquids are believed to have been used. These included gasoline, camp stove fuel, and kerosene. The specific mixing and placement of the accelerants is not known. The sets were established and the accelerants were poured between 0730 and 0830 hours, according to participants. The sets included a flammable liquid pour on the kitchen floor, papers in the partially open oven, and a combination of papers and flammable liquids on the kitchen counters. Waxed paper sprinkled with gasoline was used as a trailer to extend the fire to additional liquid pools in the adjoining rooms and upstairs to the bedrooms. In the ground floor area, sets included rubber gloves full of flammable liquids suspended over candles and couches that had been soaked in accelerants and covered with readily combustible materials. On the second floor, accelerants had been poured on clothes in partially opened drawers and on the floor and a one gallon plastic container had been partially filled with a flammable liquid with a fuse inserted to ignite it. Newspapers were strung along a rope to extend the fire across a bedroom. (See Appendix B for the layout of the sets.)

The drill participants began arriving at the site at approximately 0800 hours. Most of the participants toured the house to examine the arson sets and several later commented on the strong smell of hydrocarbon fuels and the visible puddles in several areas.

The Fire Chief and the Captain of the Milford Fire Department directed the establishment of attack hoselines and exposure protection lines. These were supplied by portable drafting tanks and the chief of a participating mutual aid department was assigned to establish the tanker relay. Two 2-1/2 inch and two 1-1/2 inch hoselines were prepared, with the larger lines positioned for exposure protection. A 1-1/2 inch line was placed near the single doorway.

All lines were charged and ready to operate. The Explorer Scouts, who were to operate the 2-1/2 inch line on the east side, were instructed on its use and allowed to flow it briefly for practice.

FIRE SEQUENCE

The fires were ignited at approximately 0845 hours. It appears that the original plan was to ignite one or more of the arson scenarios on the ground floor level and to allow the "trailers" to extend the fire to other sets on the first and second floors. The assistant chief and a firefighter entered the house to ignite the fires and to take pictures.

When the first fires were ignited, they burned with much less than the anticipated intensity and failed to spread via the "trailers." Photographs taken by the assistant chief show several independent fires burning at the different points of origin and moderate smoke conditions inside the ground floor. After several minutes the assistant chief and the firefighter exited from the house to obtain an additional fuse and to reload the camera.

The assistant chief and a different firefighter reentered the house and went to the second floor to ignite the fires on that level. Both were wearing full protective clothing and SCBA and they noted only a small amount of fire on the ground floor as they passed by.

The captain had assigned a crew to make the interior attack after the fires were burning sufficiently to produce the desired results. This attack crew included four individuals, selected to give them additional experience in interior operations. They were standing by outside the door with a charged 1-1/2 inch hoseline, all wearing protective clothing and SCBA. Some or all of these individuals may have been using air from their SCBAs while waiting.

With the delay, the captain directed this crew to go inside and observe the fire conditions, apparently to see how the fire conditions were developing. The four trainees left their hoseline outside, entered and walked through the ground floor, noting a small amount of visible flame in the kitchen and dining area. They proceeded to the second floor where they encountered the assistant chief and the firefighter who were in the process of igniting the fire in the south bedroom.

At some time in this sequence, an observer was asked to break the two windows on the south side of the second floor. This was intended to provide ventilation to help the fires to burn more freely. The windows were broken by throwing rocks from the outside.

After several minutes inside the structure, the six members on the second floor noted a sudden increase in smoke conditions and attempted to leave via the stairs. At the bottom of the stairs they found the dining room area fully involved and retreated back up the stairs. The visibility rapidly deteriorated to zero and they encountered fire coming up the stairs, as well as from the south end of the second floor corridor and through holes in the ceiling. The firefighters lost contact with each other as they searched for an exit.

Three of the trapped members managed to escape through the window adjacent to the top of the stairway, onto the kitchen roof. The assistant chief was the first out the window, followed by one of the trainees and the firefighter who was helping to ignite the fires. By the time the third member came out of the window, flames were coming from the top of the opening and he received burns on his hand and arm as he hung from the window sill, before dropping to the ground. The third firefighter's SCBA was damaged by the heat, resulting in a rupture of the low pressure hose and melting of straps.

The members outside the structure were unaware that the firefighters were trapped for a period of time. They had noted a rapid increase in smoke production and fire intensity, followed by failure of the two ground floor windows at the south end of the house. When the second floor window was broken open from the inside and the assistant chief crawled out, the members outside ran to get a ladder to assist him and the other two escapees down from the kitchen roof.

A few moments later, the front window of the north bedroom on the second floor was broken out and witnesses reported that one of the trapped firefighters attempted to crawl out through the opening. His SCBA cylinder would not clear the opening and he withdrew back into the smoke. He appeared at the opening a second and possibly a third time, struggling to escape, before disappearing back into the smoke. The witnesses reported that his SCBA facepiece was missing the last he was seen.

Ladders were raised to this window and repeated attempts were made by several members, including the assistant chief, to enter and remove this victim. He was finally located and removed, but efforts to revive him were unsuccessful and he was dead on arrival at the regional hospital. Estimates of the time he was trapped vary from 5 to 15 minutes. During the rescue efforts the assistant chief was overcome by smoke, and was also transported to the hospital.

Requests for additional mutual aid assistance were transmitted by radio and several additional volunteer companies responded from surrounding areas. The standby hoselines were operated into windows in an attempt to knock down the fire which eventually vented through the roof at the south and at the attic. With the assistance of the mutual aid companies the fire was controlled sufficiently for crews to enter and search for the two missing members.

The bodies of the two remaining members were found in the corridor, near the stair opening. Both were removed and transported to the hospital, but they too were dead on arrival. Witnesses reported that one or both of these victims still had their SCBA facepieces in place when found, but they had become separated from their air cylinders and backpack harnesses.

CAUSE OF DEATH

The cause of death for all three victims was listed as inhalation of products of combustion. The victim who tried to escape via the bedroom window had a reported carboxyhemoglobin level of 27.4 percent. The two victims in the corridor had levels of 48.2 percent and 54.8 percent. All three had soot and particulate in their breathing passages, supporting the conclusion that they died of smoke inhalation. More extensive blood analysis to test for levels of other toxic products of combustion was not carried out.

All three victims also had varying degrees of burns, but the autopsy report indicates that these were insufficient to cause death or occurred after the victims were deceased. The victims were in the fire environment for an extended time before being removed.

OTHER INJURIES

The assistant fire chief was treated for smoke inhalation after his SCBA ran out of air during the attempted rescue operations.

The firefighter who was assisting in lighting the fires sustained burns to his hand while hanging from the window sill and was also treated for smoke inhalation. He reported that the breathing tube of this SCBA failed as he escaped through the window.

The surviving trainee received burns to the side of his face when he pulled his facepiece away from his face in order to talk while still inside the house.

The assistant chief and the trainee were treated and released from the hospital. The other firefighter was admitted to the area burn center for treatment and released several days later.

ANALYSIS

Many different factors contributed to this incident, each making the situation more dangerous and all having a significant influence on the tragic results. These factors include errors in judgment, as well as physical factors that relate to the behavior of the fire under these particular circumstances.

The fact that several fires were set in the house and liberal use was made of accelerants is a major factor. When the fires were first ignited, the accelerants may have been present for an hour or more, providing ample time for vapors to saturate the atmosphere and for liquids to soak in to other fuels. With only one door at the most remote corner of the structure the ventilation was extremely limited and the vapors may have displaced much of the available oxygen. These circumstances may have created an atmosphere that was fuel-rich and oxygen-deficient, explaining the failure of the individual fires to burn freely and extend rapidly.

Based on the amount of accelerants involved, it could also have been possible to create an atmosphere in the house that was within the flammable limits of the accelerants. Under those circumstances, the house could have become a "bomb" and exploded violently when the first source of ignition was introduced. There would be no way to predict either of these circumstances without test instruments and/or extensive calculations.

Ignition of some of the sets may also have been inhibited by the water that leaked in when the Explorer Scouts were being trained with the 2-1/2 inch hoseline.

The flashover of the rooms on the ground floor was very predictable, based on the amounts and arrangements of combustibles that were present. The delayed flashover is consistent with a ventilation-controlled fire and the breaking of the upper windows may have provided sufficient air flow to trigger flashover when fresh oxygen was introduced through the open door.

In addition to multiple fires heating the atmosphere and accelerants adding to the fuel load, the interior finish materials on the ground floor are recognized as contributing to rapid intense fire growth. Low density fiber board tile and wood wall paneling have been noted as contributors to many other tragic fires, including fatal training fires.

When the fires on the ground floor reached flashover, the members on the upper level was trapped because their only exit was through the involved area. Examination of the fire scene reveals that the fire extended through voids in the floor and walls in the south east corner, into the former bathroom area and into the attic. This exposed the firefighters to flames in the corridor and coming down through openings in the ceiling, as well as up the stairs. Holes in the walls and ceilings allowed the fire to extend and to come back down on the victims from above.

It is difficult to determine exactly what happened to the three victims when they became trapped. The survivors reported zero visibility and intense heat that obviously made escape difficult. The fact that two of the victims were found within a short distance of the window that provided escape for three others suggests that they might have escaped if their protective equipment had provided more time or if they had been better trained or more experienced in how to escape from a critical situation. The short time that is available for escape within the limitations of protective clothing and equipment means that all firefighters must be trained to react to flashover and to have an escape route.

All of the firefighters were reported to have been using SCBA at the time they became trapped. The fact that the cause of death of all three victims was carbon monoxide inhalation provides strong evidence that failure of the breathing apparatus was a direct factor in preventing their escape. This is reinforced by the reports that the SCBA worn by the last firefighter to escape suffered a rupture of the low pressure hose and melting of straps as he escaped through the window. The circumstances do not suggest that running out of air was a significant probability, due to the relatively short time sequence.

The inhalation of carbon monoxide would have quickly incapacitated the victims at the levels that would be expected inside the house. This inhalation could have occurred when SCBA components failed, such as hoses, regulator diaphragms, exhalation valves, lenses or lens retaining devices. It could also occur when the facepiece seal is compromised due to (1) an impact (such as falling), (2) being pulled loose when strap failure caused the air cylinder and backplate to fall off the victims back, or (3) the victim pulling off the facepiece to call for assistance.

It is not clear, specifically, which of the victims still had the facepiece in place when found and/or why the other(s) may have loosened or lost the facepiece seal. One SCBA facepiece had a missing lens, one had a broken lens, and at least two had severe damage to the low pressure hoses, any or all of which would have negated the protection offered by the SCBA.

The melting of one or both shoulder straps would result in dropping of the SCBA air cylinder and backplate. This could result in mechanical failure of the low pressure hose, pulling loose the facepiece seal, or both. The examination of physical evidence strongly suggests this possibility. The loss of the air cylinder during a flashover would probably be a fatal situation and is strongly suggested in this incident.

It appears that the victims were incapacitated by inhalation of carbon monoxide and quickly rendered unconscious. Their burn injuries do not appear to have been sufficient to cause death or rapid incapacitation and the majority of the burns may have occurred after the victims were overcome by carbon monoxide inhalation. This suggests that respiratory protection was the weak link and that protective clothing provided sufficient thermal insulation to allow for additional escape time.

COMPARISON WITH NFPA 1403

This training exercise deviated from the safety standards of NFPA 1403, "Standard for Live Fire Training Evolutions in Structures," in several significant areas. There is strong evidence to suggest that this tragic event would have been avoided if the requirements of the standard had been applied. It must be noted that there is no legal requirement for compliance with NFPA 1403 in the State of Michigan. The standard would prohibit the use of flammable liquids and restrict the use of combustible liquids to a small amount to assist in ignition.

NFPA 1403 specifically points out the dangers of highly combustible interior finish materials and would specifically require removal of the ceiling tile and possibly the wall paneling on the ground floor before using the house for live fire. The holes in walls and ceilings would have required repairs or coverings and a railing around the stair opening would have been required. The standard would also require the removal of the trash and debris from the house.

The standard also calls for roof openings to be provided in advance to provide for emergency ventilation.

A pre-burn briefing and familiarization safety tour for all participants would have been required prior to igniting the fires and a safety officer would have to be assigned to monitor the safety of the entire operation. The safety officer would have to approve the ignition sets, as well as monitoring the safety of all participants.

One qualified instructor is required to be with each group of trainees and a fireground organization would have to be in effect to maintain accountability for the trainees and all other involved persons.

A building evacuation signal and plan would have to be established and emergency medical personnel would have to be standing by at the scene before a drill could begin.

In this case, there were no specific assignments made in advance and no fireground organization was in place to guide the operation. Instead, it appears that several individuals were responsible for various parts of the operation, but no overall command structure was provided. Communications and coordination were severely lacking before during, and after the ignition of the fires and after the members became trapped.

The arrangement of this particular structure, with only one exit located in a remote corner of the structure on the ground level, constitutes additional hazard. It would be a function of the planning and preparation team and the safety officer to identify this hazard and to provide alternate means of escape or to prohibit interior operations in this structure.

LESSONS LEARNED

It is unfortunate that the lessons learned from this particular incident only reinforce training lessons that have been learned previously. The development and adoption of NPFA Standard 1403 was a direct result of earlier incidents which involved many of the same factors and resulted in deaths and injuries to firefighters and trainees.

1. **Lack of Awareness of Past Training Lessons and Standards**

    The most significant lesson to be learned from this incident may be that the fire service in many areas continues to be unaware of NFPA 1403 and other safety standards and experiences that have been well documented in fire service publications, training programs, and conferences. There continues to be a need to reach out to the fire service in large and small communities to disseminate this information and there is an equal need for responsible individuals in the fire service to make themselves aware of critical information that is available on training safety.

    While it is easy to suggest that errors in judgment occurred in this incident, this only reinforces the observation that individuals responsible for training safety are not receiving the information that they need for the protection of their trainees.

2. **Potentially Hazardous Fire Scenarios for Training**

    The dangers involved with the use of flammable liquids, highly combustible interior finish, and multiple points of origin in training are very evident in this situation.

3. **Potentially Hazardous Operating Procedures for Training**

    Training should address the danger that is involved when members enter structures without a charged hoseline, go above a fire, operate without supervision, and are not part of an overall fireground accountability system.

4. **Training for Escape**

    The reaction of the trapped firefighters to the rapidly deteriorating conditions inside the house may reflect their level of training and experience in dealing with critical conditions. Training must focus on how to escape when fire conditions change suddenly and unexpectedly, whether it be training or operational situations.

5. **Protective Clothing Works**

    Protective clothing that meets NFPA standards provides good protection from critical fire exposure, at least to prevent serious injury while escaping, if escape can be accomplished within a reasonable time. The condition of clothing worn by both the survivors and those who did not survive this fire suggests that good thermal protection was available for several minutes in this fire. That was not the problem here.

6. **Vulnerability of SCBA Units in Flashover**

    There is evidence that premature failure of components of their SCBA may have been a critical factor in preventing the escape of the victims. Further analysis is needed to determine if inherent weaknesses in the design and construction of SCBA constitutes an unidentified danger to firefighters in critical flashover and fire exposure situations.

The fire resistance requirements for some SCBA components, particularly belts and straps, have been increased in the last year and several manufacturers have changed the construction of particular components to meet these standards. But there still are many SCBA units in service which do not meet the new standards, and there may be additional components, such as hoses, facepieces and regulators, which have not been adequately tested under fire conditions.

APPENDIX A

Observations on Protective Clothing and Equipment

This appendix documents observations and photographs of clothing and protective equipment worn by the injured and deceased firefighters in the Milford fire. The items listed were retrieved and identified as having been worn by the individuals noted. Not all of the clothing and equipment was retrieved or identified; the omission of an item below does not mean it was not worn, just that it was not retrieved.

All of the individuals involved in the fire were wearing protective clothing that was essentially in compliance with NFPA standards, and they all used SCBA. Some of the firefighters wore 3/4-length rubber boots and one or more wore knee-length rubber boots and protective trousers.

The individuals wore a variety of civilian clothing under their protective clothing. While this clothing did not meet the fire resistant performance standards, that would apply to station uniform clothing according to NFPA 1975, there is no evidence to suggest that the clothing had any positive or negative impact on burn injuries.

This information should be of value to individuals who are involved in research and development of protective clothing and equipment.

DALE WILTSE
Milford, VFD   Suffered hand burn and smoke inhalation.

COAT: Black Nomex outer shell with green needlepunch Nomex liner bonded to black neoprene moisture barrier.

Label: Janesville, Date: 4/3/84, Req: 578400, Name: 5746. Meets NFPA 1971.

Coat was intact with some damage to reflective trim (Scotchlite) and some charring to lower right side of outer shell. Minor charring to bottom of inner liner material.

BOOTS: Short, rubber – no damage noted.

OTHER CLOTHING: Nylon windbreaker Tee shirt: Hanes Beffy Tee 100 Percent Cotton. No fire damage noted to either item. Gloves not retrieved.

PHOTOS: 1, 2, 3

ROBERT A. GREGORY
Highland TWP VFD   Deceased due to inhalation of products of combustion. Found in bedroom.

WORKSHIRT & PANTS: 65 percent Polyester/35 Percent Cotton, dark blue, uniform type – no fire damage noted.

COAT: Nomex outer shell, needlepunch Nomex bonded neoprene liner.

Label: Morning Pride – 100 Percent Nomex. Some char damage to outer shell. No damage to liner noted. Parts of SCBA waist strap fused to outer shell on left side.

GLOVES: (Right only) – lined rubber (no label). Fire damage to rubber.

HOOD: Tempo Uno – knit Nomex/Wool. Wool appears to have been worn outside and is approximately 50 percent charred or burned. Nomex is intact.

SCBA: Scott Air Pak II – Pressure Demand switch on, Main Line Valve on, Bypass Valve off. Part of waist strap attached to back plate has been cut off. All other straps melted off. Facepiece lens is missing with some fire damage noted on retaining ring only. Breathing tube is intact.

PHOTOS: 4, 5, 6, 7, 8, 9, 10, 11, 12, 13, 14

THOMAS B. PHELPS
Lyon TWP VFD   Deceased. Found in corridor adjacent to window.

COAT: Nomex outer shell, needlepunch Nomex liner bonded to Neoprene moisture barrier.

Label: "meetings NPFA 1971."

Coat is heavily damaged by fire exposure, particularly back and bottom half of sides. SCBA straps and buckle fused to outer shell.

GLOVES: Lined rubber – some damage to rubber.

BOOTS: 3/4 length rubber – rubber is soft and tacky, but boots are intact.

PANTS: 65 Percent Polyester/35 Percent Cotton – blue, uniform type. Domestic Linen Supply. Charred above 3/4 length boot line.

OTHER CLOTHING: Underwear and tee shirt: 50/50 Cotton/Polyester. Cotton (100 percent) flannel sport shirt. All items charred below belt level and several inches above. Consistent with exposure above boots and under coat.

SCBA: Scott Facepiece, soot covered inside and outside. Breathing tube is torn loose, approximately 12 inches from facepiece. No fire damage evident to facepiece.

HELMET: Bullard – faceshield severely melted and deformed. Outer shell partially melted. Damage to impact liner is not evident (not removed).

PHOTOS: 15, 16, 17, 18, 19, 20, 21, 22, 23, 24

MARSHA BACZYNSKI
Milford TWP VFD

	Deceased. Found with foot hanging over stair opening.
COAT:	Janesville with NFPA 1971 label. Nomex outer shell, quilted Nomex liner with Neoprene moisture barrier. Coat is essentially destroyed; most of liner is intact.
PANTS:	Janesville: date: 12/16/86; Req: 439100, name 13558. Nomex outer shell with needlepunch Nomex liner, bonded to Neoprene vapor barrier. Pants are essentially destroyed – most of liner is intact. Waist area and suspenders undamaged.
HOOD:	Charcate 2-ply Nomex. Some charring noted, but intact.
HELMET:	Cairns 660, Polycarbonate shell. Shell is melted out of shape. Impact cap appears undamaged.
BOOTS:	Short rubber type. Left almost destroyed by fire exposure. Right has minor damage.
SWEAT PANTS:	Partially burned – no evidence of drip or melt. No label identified.
SHIRT:	50/50 Cotton/Polyester (NFPA Shirt) – no fire damage evident.
BRA:	Synthetic materials – no fire damage evident.
SCBA:	Scott – lens broken and part missing. Heavily soot covered, inside and outside. Breathing tube destroyed by fire.
PHOTOS:	25, 26, 17, 2, 8, 2, 9, 3, 0, 31, 32, 33, 34, 35, 36
OTHER PHOTOS:	#37 – Remnants of unidentified melted helmet, fused to buckles from SCBA straps, shown next to impact cap from Baczynski articles.
	#38 – Items worn by Firefighter Goodnough, Milford TWP VFD.

Appendix A

continued

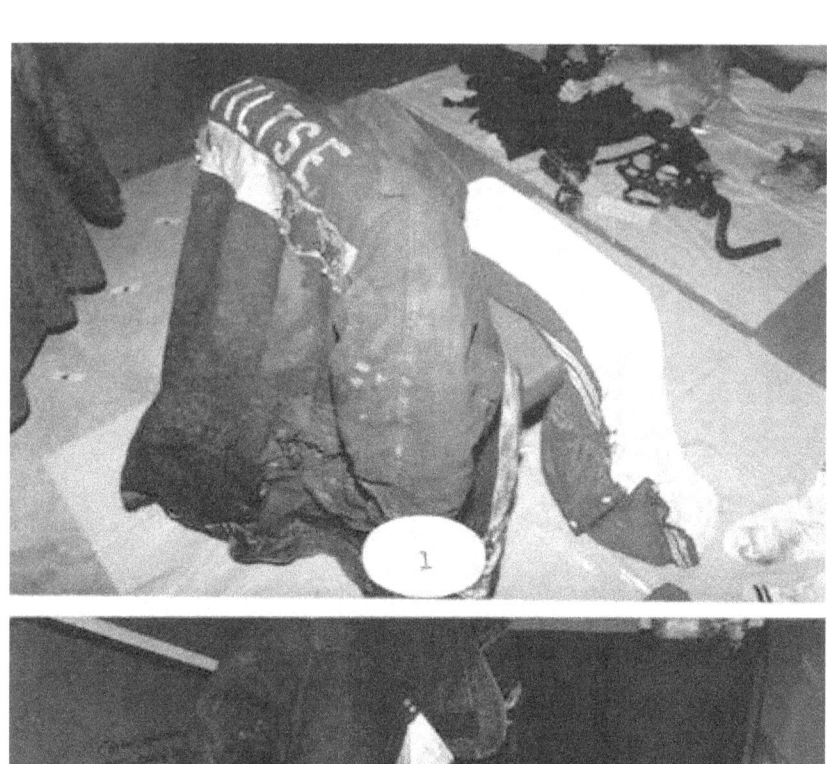

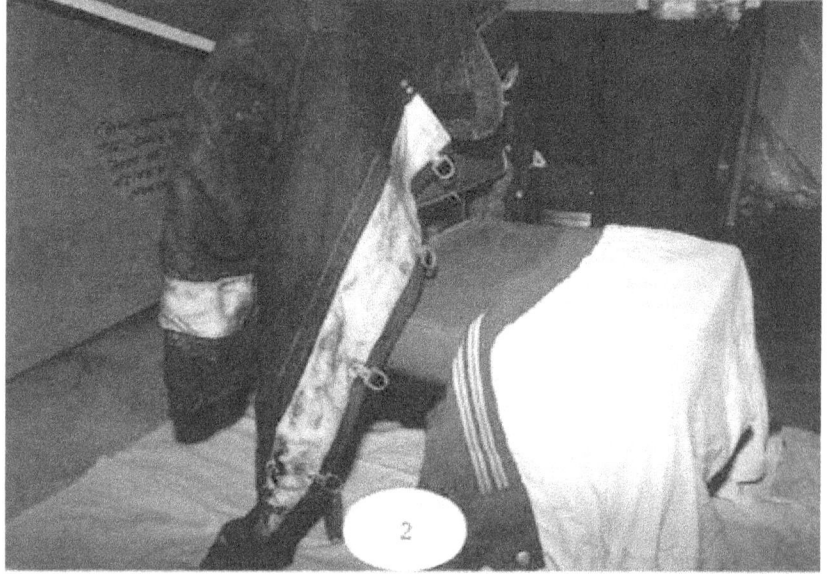

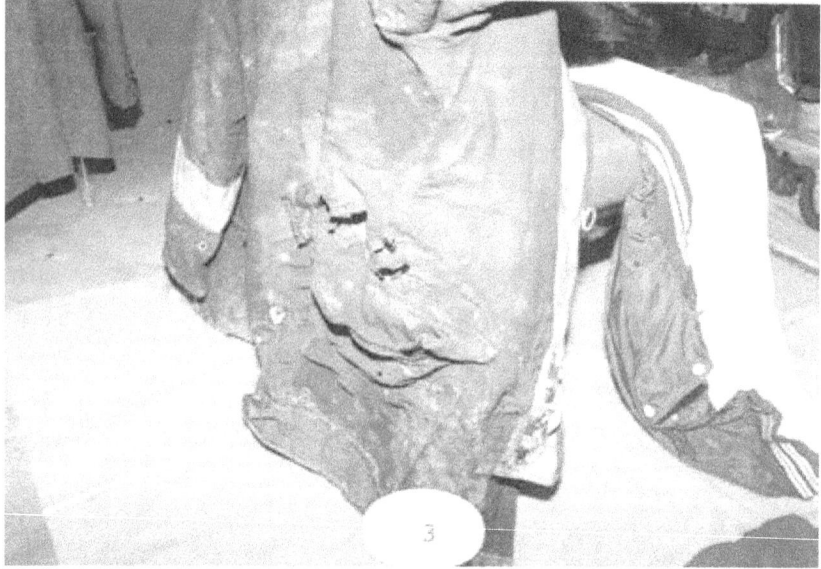

Appendix A
continued

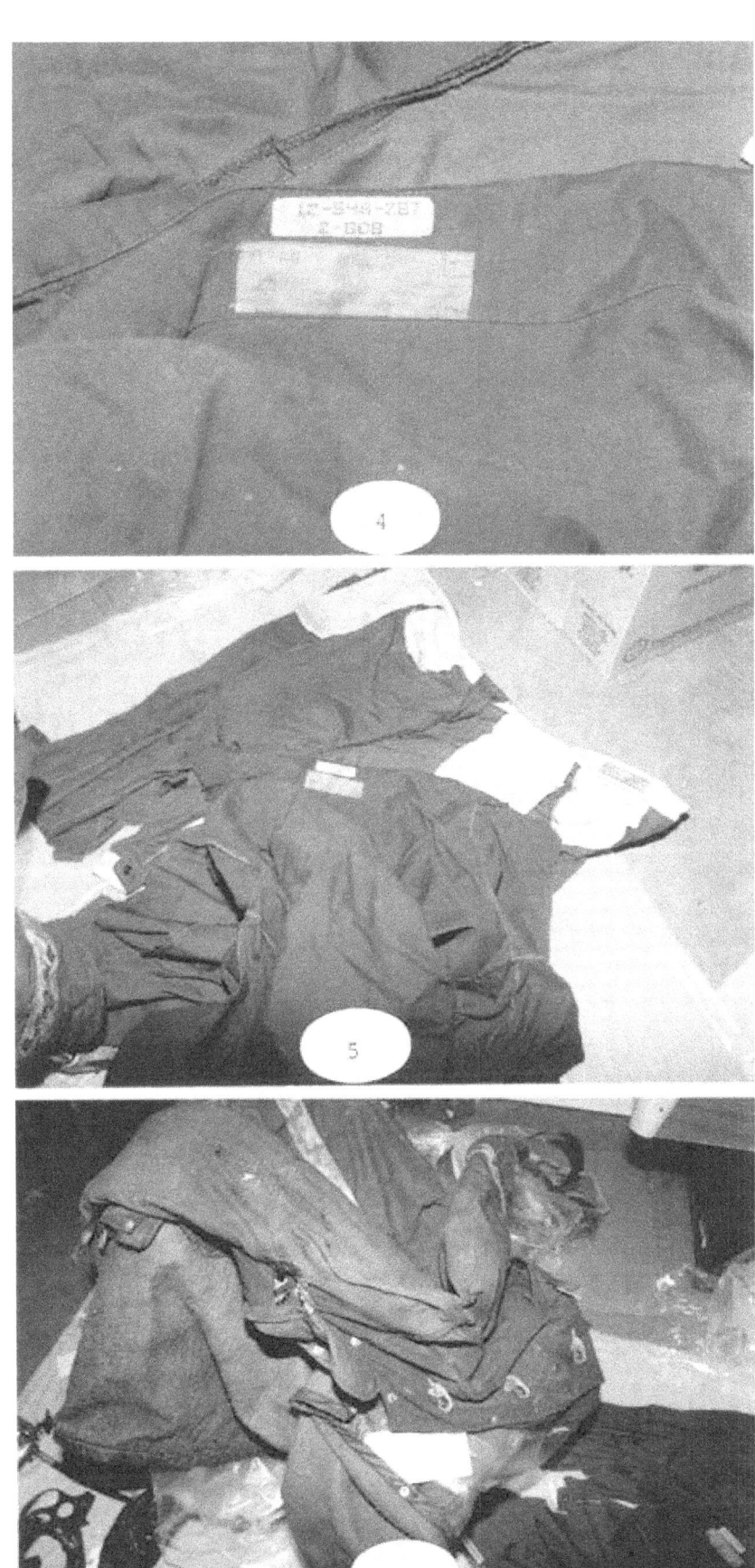

Appendix A
continued

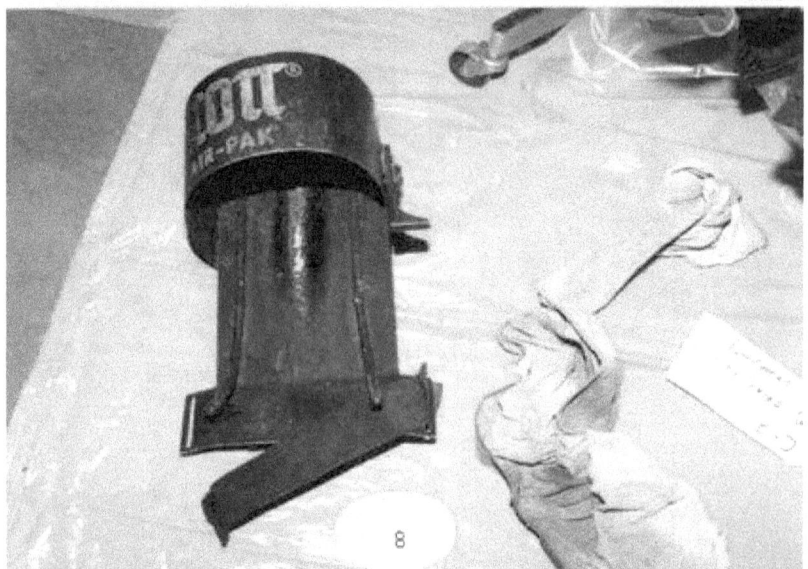

Appendix A
continued

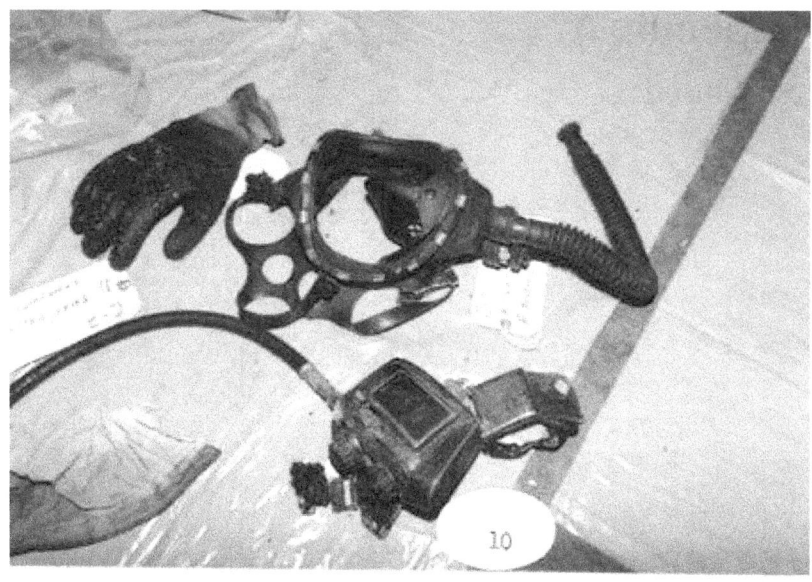

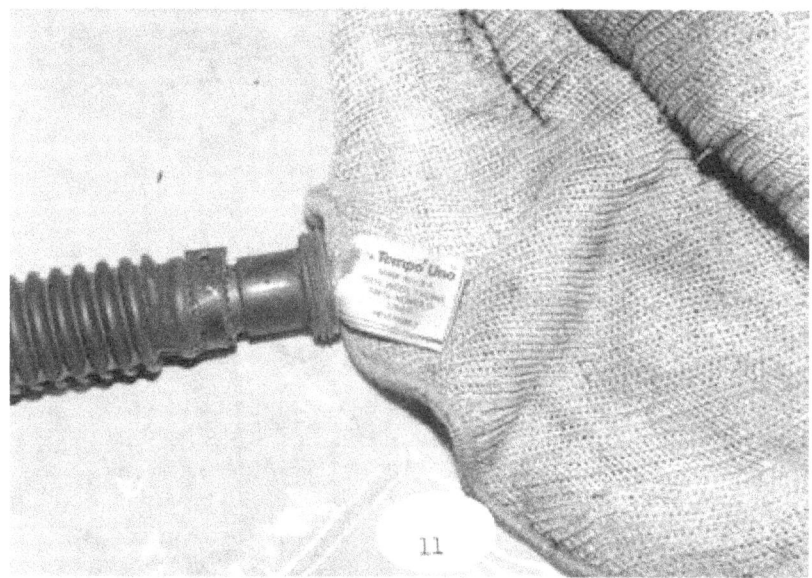

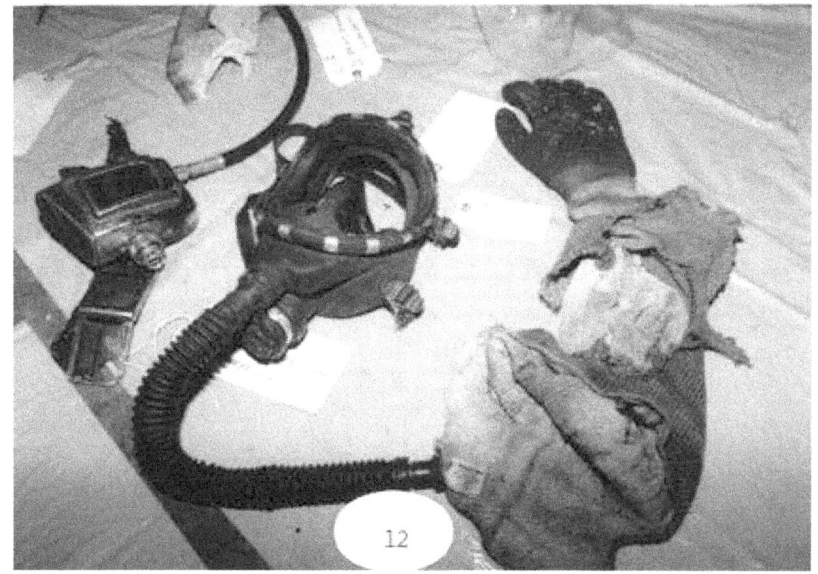

Appendix A

continued

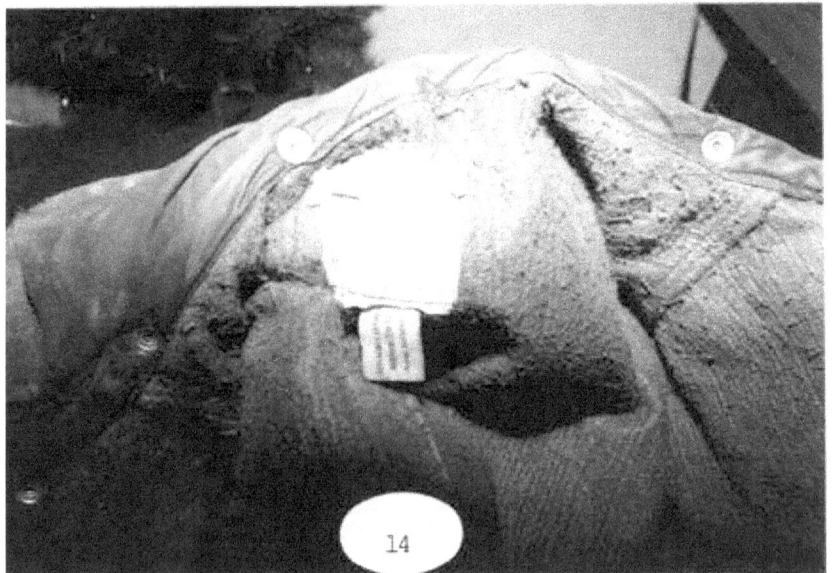

Appendix A
continued

Appendix A

continued

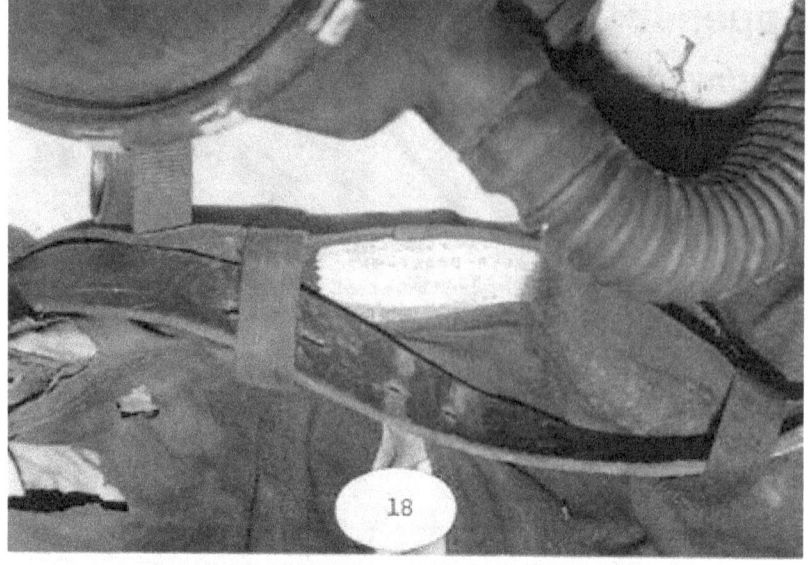

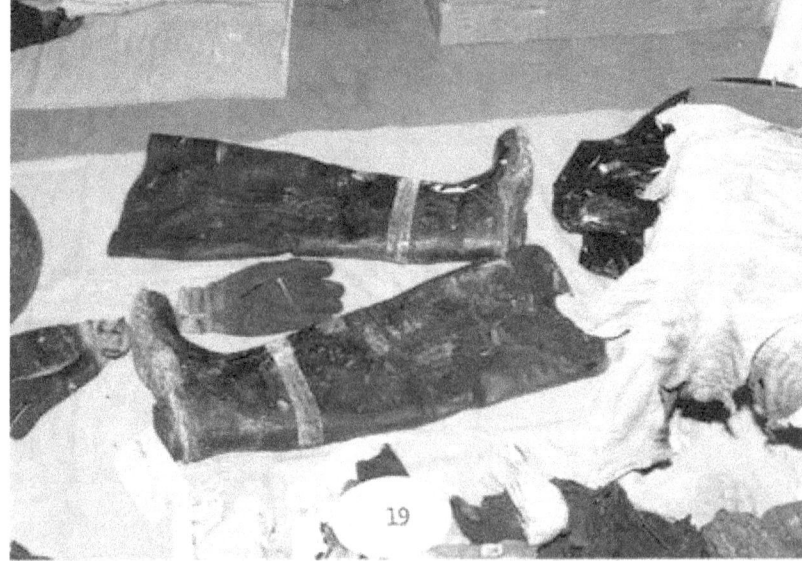

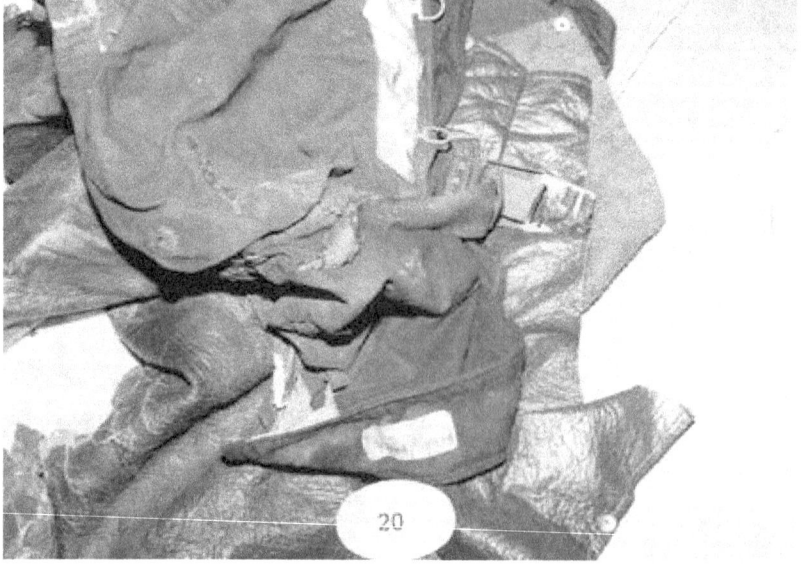

Appendix A
continued

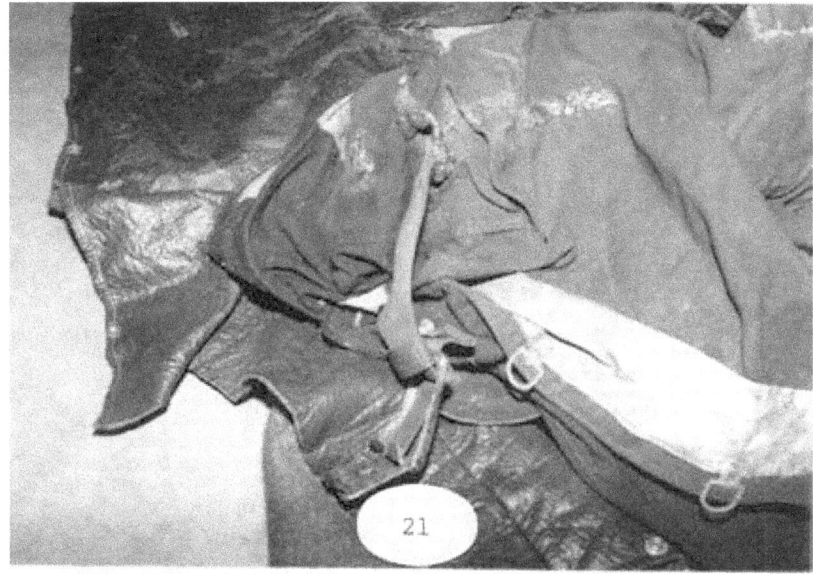

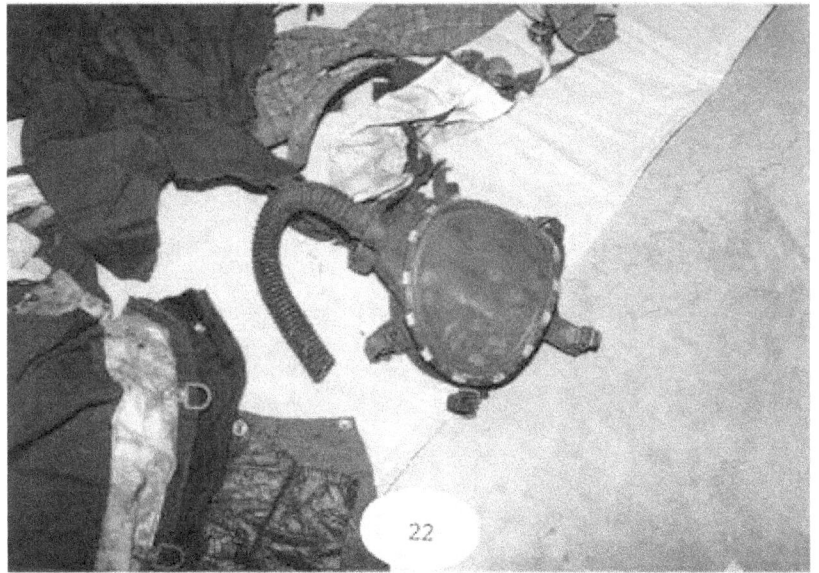

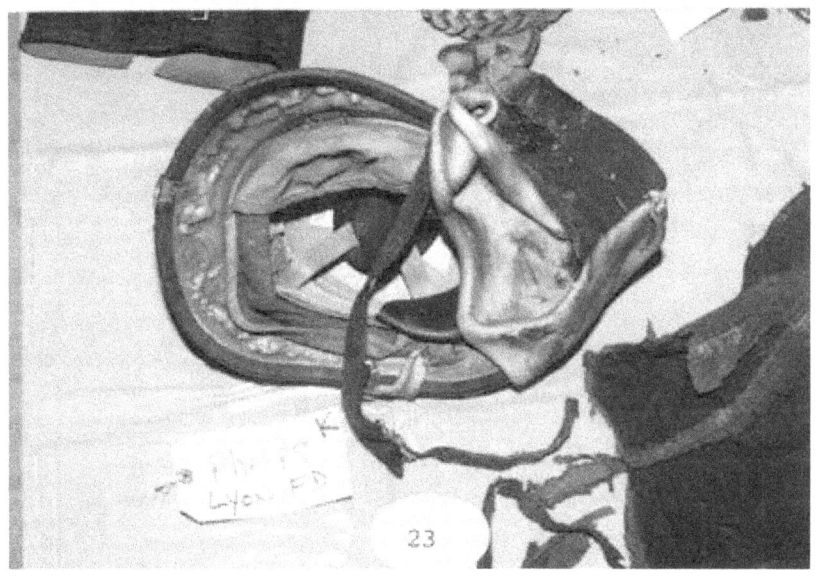

Appendix A
continued

Appendix A
continued

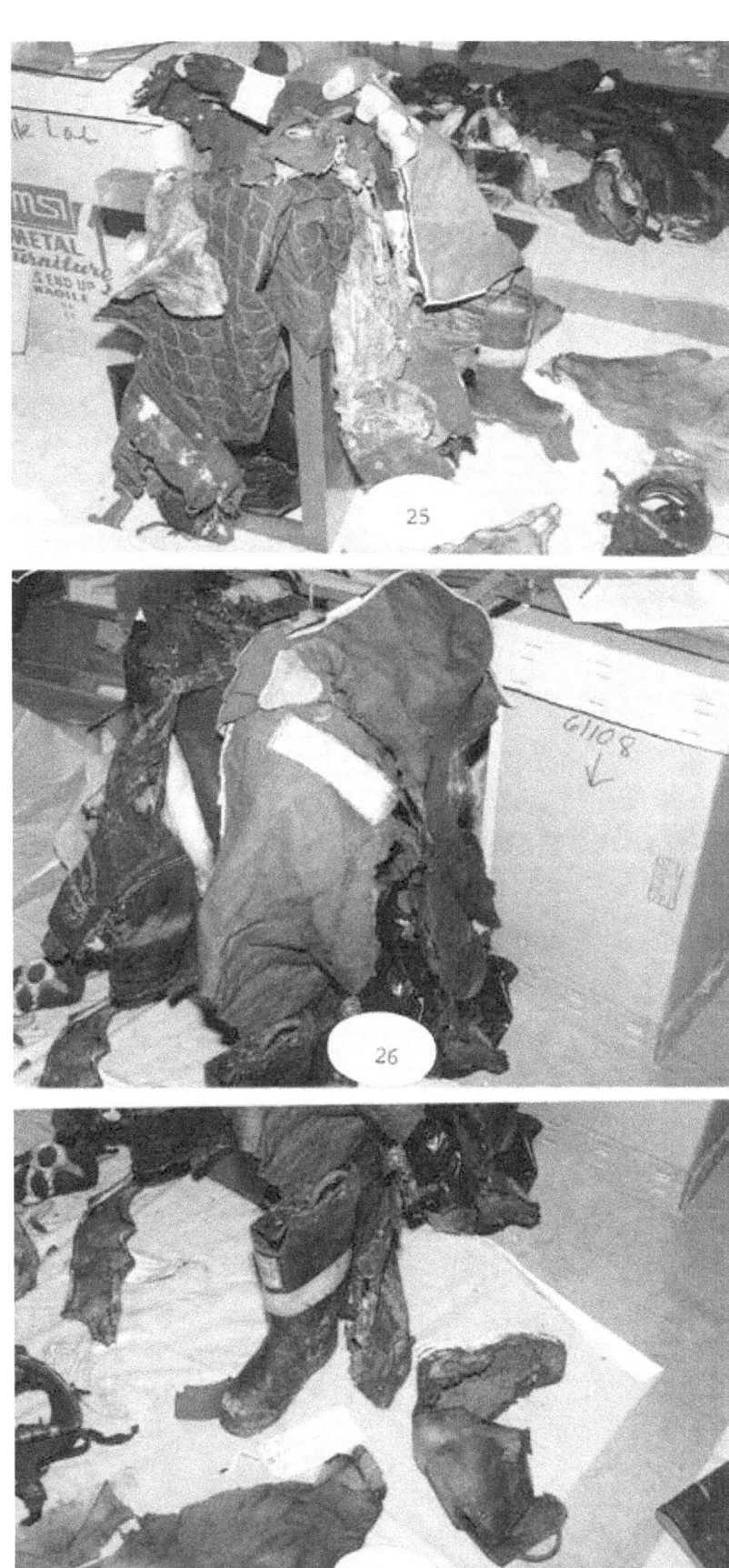

Appendix A
continued

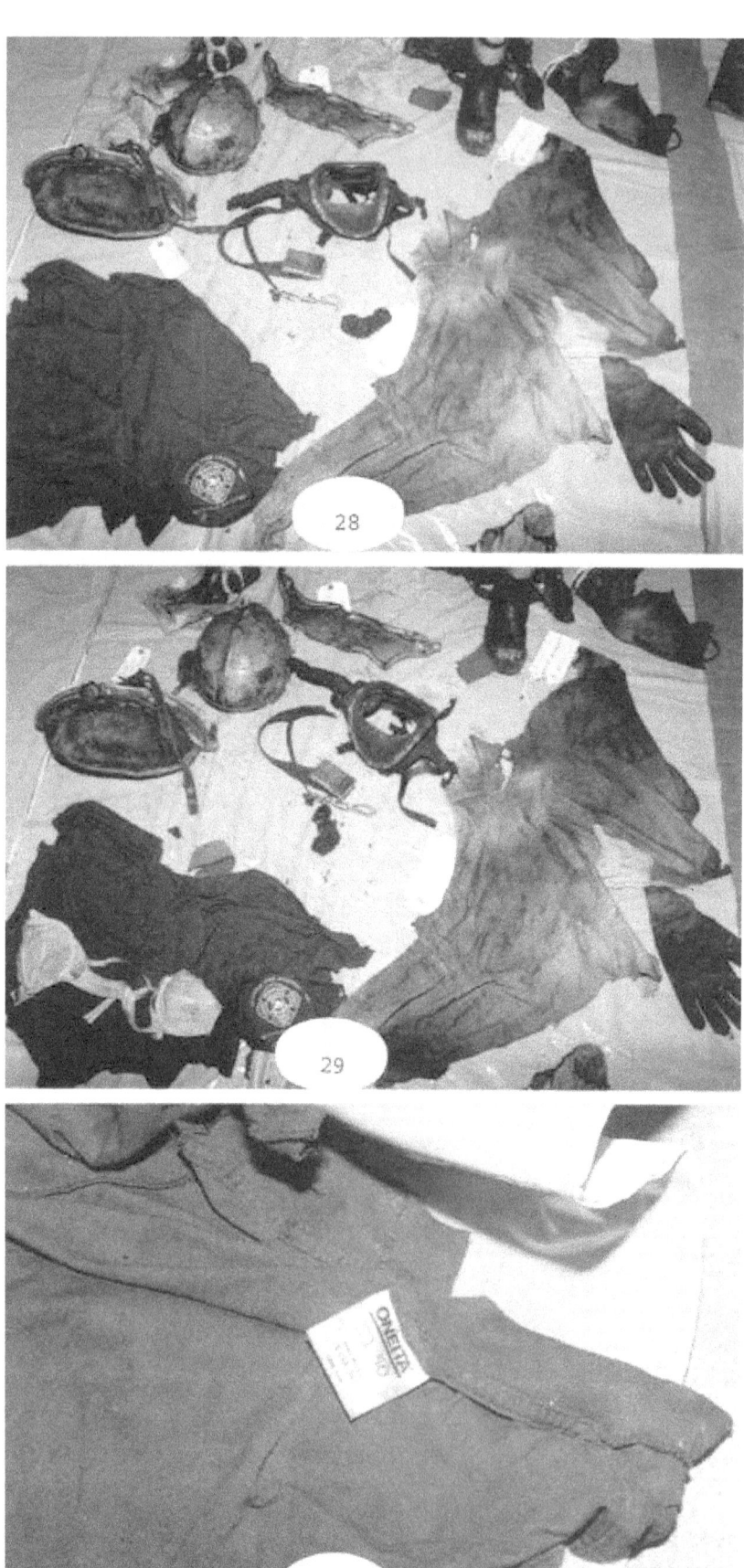

Appendix A
continued

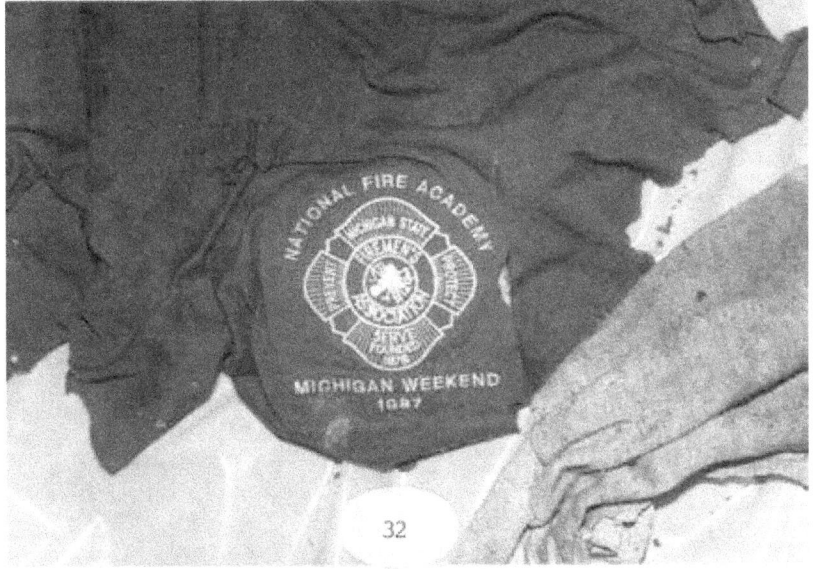

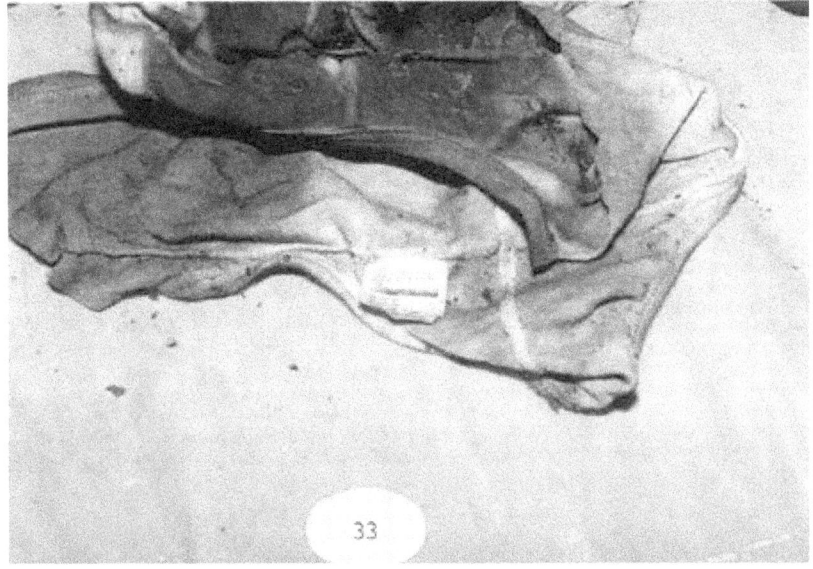

Appendix A
continued

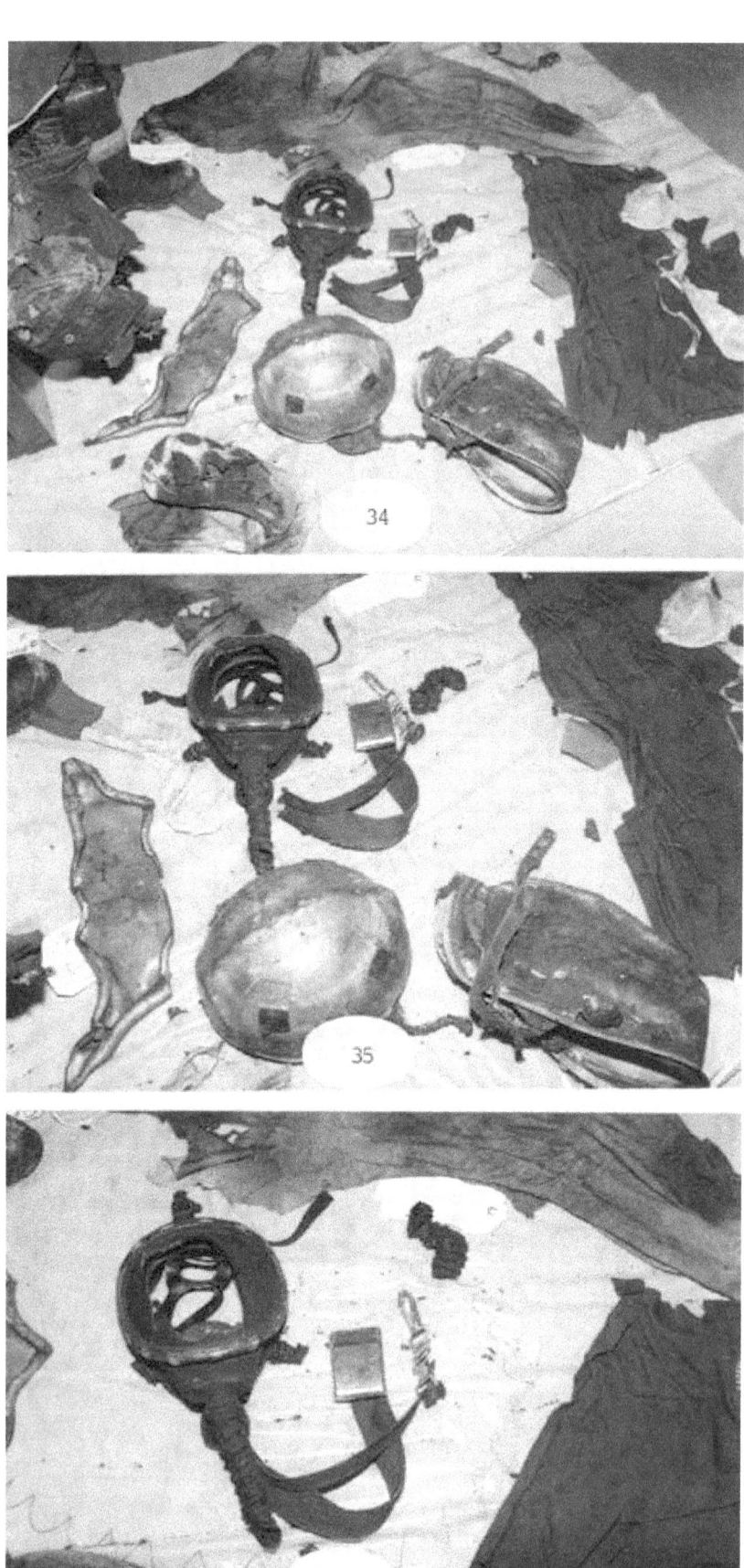

APPENDIX B

Photographs of Fire Scene and Floor Plans Showing Points of Origin

The following are photographs of the farmhouse used in the Milford training fire, fire damage, and accelerant containers. Also presented are two diagrams showing the floor plans of the house and the point of origin for each arson set listed below. In addition, there were several puddles of accelerants (flammable pours) and trails of accelerant connecting several of the points of origin.

1. Papers in oven (kitchen)

2. Accelerants and papers on "kitchen counter"

3. Rubber glove filled with flammable liquid

4. Rubber glove filled with flammable liquid

5. Couch with accelerant pour

6. Couch with accelerant pour

7. Plastic jug containing accelerant

8. Accelerant poured on clothing

9. Clothing in open dresser drawers accelerant pour

10. Accelerant-soaked newspapers on string

Appendix B
continued

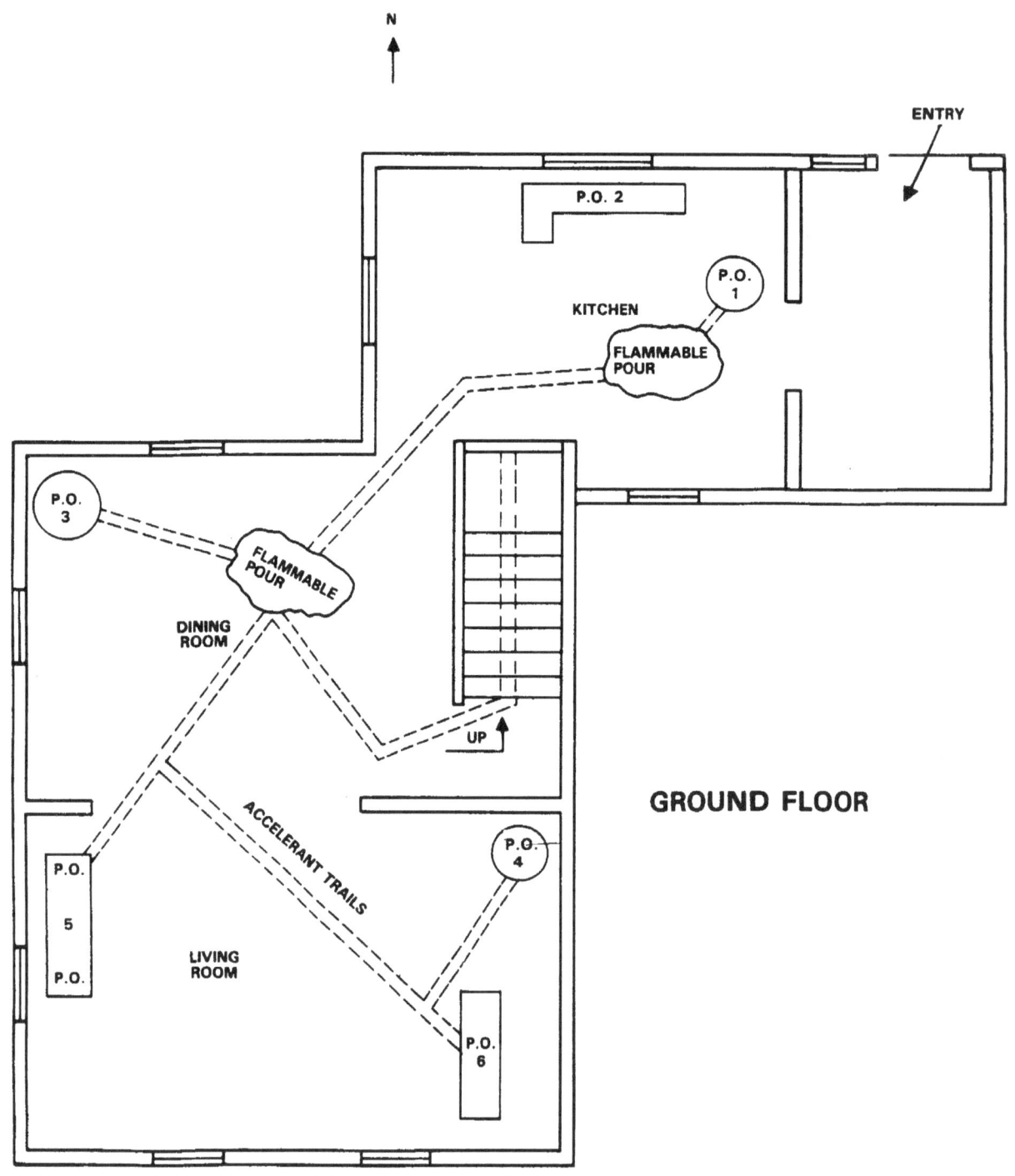

Appendix B

continued

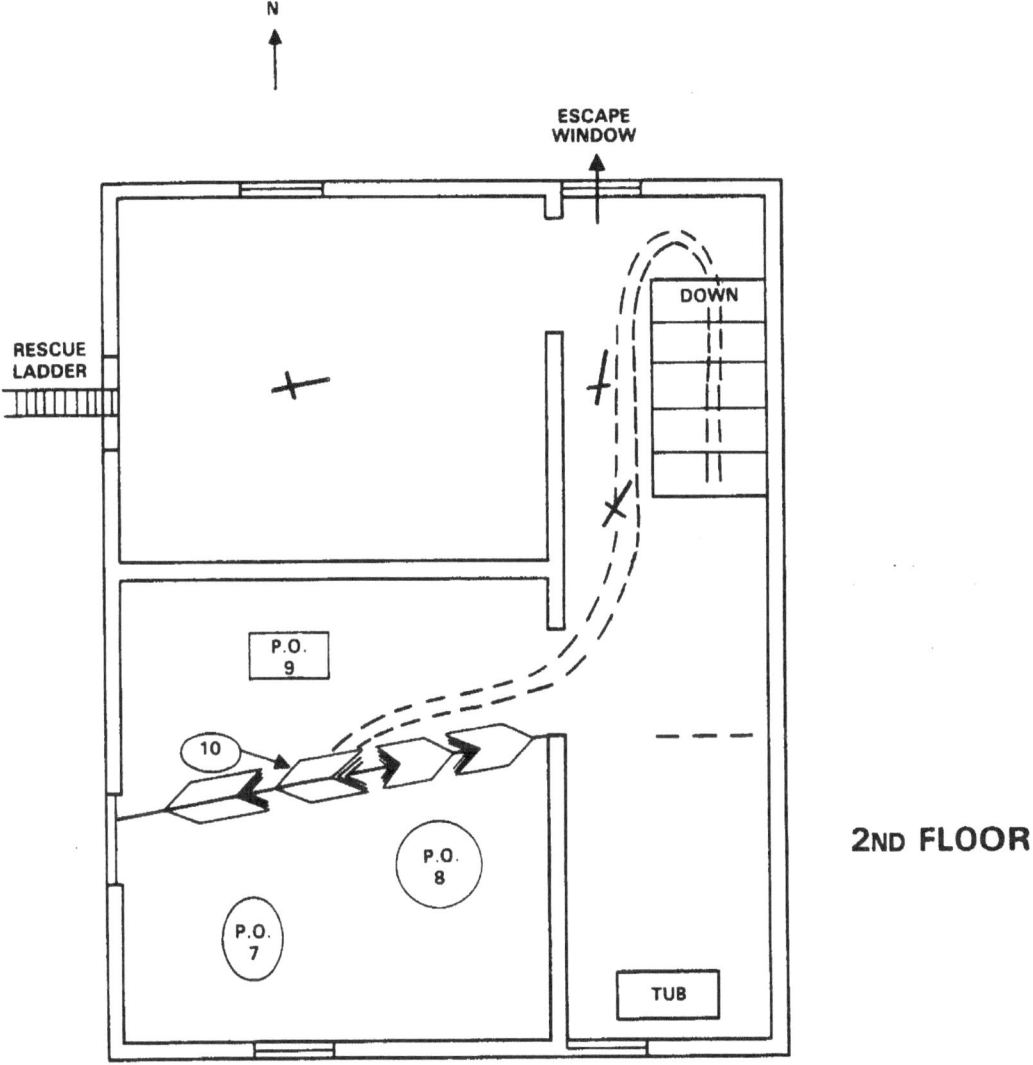

www.ingramcontent.com/pod-product-compliance
Lightning Source LLC
Chambersburg PA
CBHW081243170526
45165CB00009B/3173